SILLY BODY SCIENCE

by Robin Twiddy

designed & illustrated by Amy Li

LERNER PUBLICATIONS ◆ MINNEAPOLIS

First American edition published in 2021 by Lerner Publishing Group, Inc.

Editor: Madeline Tyler
Design: Amy Li

Lerner Publications Company
An imprint of Lerner Publishing Group, Inc.
241 First Avenue North
Minneapolis, MN 55401 USA

For reading levels and more information, look up this title at www.lernerbooks.com.

Main body text set in Mikado. Typeface provided by HVD.

Library of Congress Cataloging-in-Publication Data

Names: Twiddy, Robin, author. | Li, Amy, illustrator.
Title: Silly body science / written by Robin Twiddy ; designed & illustrated by Amy Li.
Description: First American edition. | Minneapolis: Lerner Publications , 2021. | Series: Silly science | Audience: Ages 8–12 | Audience: Grades 4–6 | Summary: "You don't have to be a scientist to discover and delight in silly (and gross) science facts about the human body, including how virtual reality tricks your brain, why we burp, and more!"—Provided by publisher.
Identifiers: LCCN 2020047896 (print) | LCCN 2020047897 (ebook) | ISBN 9781728416489 (library binding) | ISBN 9781728423890 (paperback) | ISBN 9781728418773 (ebook)
Subjects: LCSH: Human physiology—Juvenile literature. | Human anatomy—Juvenile literature.
Classification: LCC QP37 .T95 2021 (print) | LCC QP37 (ebook) | DDC 612–dc23

LC record available at https://lccn.loc.gov/2020047896
LC ebook record available at https://lccn.loc.gov/2020047897

Manufactured in the United States of America
1-48895-49209-1/22/2021

CONTENTS

Words that look like **THIS** are explained in the glossary on page 31.

SCIENCE:

Why So Serious?

We all know science, don't we? Boring people in boring white coats, writing boring things on boring clipboards. Boring, boring, boring, right? Wrong! What if I told you that science can be sillier than you have ever imagined?

Silly BODIES

Bodies are silly.

Bodies are no laughing matter. Complicated brains, disgusting illnesses, really bad farts . . . Wait, that doesn't sound very serious, does it? It turns out that bodies can be much sillier than anyone thought.

Welcome to the

SILLY ZONE

You are about to enter the Silly Zone. In this book you will find scientists doing the strangest things. Get ready!

Real

SCIENCE

Although this science all seems very silly, there is often a very good reason behind it all, or at least something very useful that we can learn from it. Scientists once got men to wear wet underwear on cold days and measured the temperature of the men's bottoms. From this, they learned that the thickness of the underwear was more important than the type of material.

It Takes GUTS to Ride a Roller Coaster

What Is INSIDE You?

You are a disgusting pile of wet, slimy organs and muscles, all held together by skin and bones. But don't worry—everyone else is too. When you are walking around, all the organs in your body are packed together while being pulled down to Earth by GRAVITY. And somehow this feels normal to us.

Let's Go to the THEME Park

Have you ever been on a roller coaster with a big drop? If you have, you probably felt a strange feeling in your body, sometimes called a stomach lurch. You may have also puked everywhere. But we're not here to think about the science of sick—let's talk about the science of the stomach lurch.

The Science of the Stomach LURCH

When you drop on a roller coaster, the force of the drop cancels out the force of gravity. That means all the organs that were being pulled down and packed together are no longer feeling that force. That's what makes your stomach lurch. It also makes you feel weightless, and sometimes sick.

This same feeling can be felt if a car goes over a hill too fast.

What Would It Feel Like to Fall FOREVER?

If you don't like this feeling on a roller coaster, then, whatever you do, don't go into space. You may think it's bad enough to feel it for a few seconds on a roller coaster, but astronauts feel something similar for the whole time that they are in **ORBIT** around Earth.

The Speed of THOUGHT

The DELAY

Wiggle your leg. Now stick your tongue out. Now smush your face up with your hands. Excellent work. When you want to move your body, it happens as soon as you think about it, doesn't it? Well, actually, it doesn't happen right away. The gap between thinking about moving and actually moving is very small, but it is there. To find it, let's look inside your brain.

Calling the BRAIN

Your body is full of special **CELLS** called neurons. Neurons send messages between your body and your brain. For example, if you touch something hot, your fingers send a message to your brain through the **NETWORK** of neurons in your body. And that message is usually "OUCH."

SIGNAL Me

When you want to move your body, your brain sends a message through the neurons to the body part that needs to move. Messages in the body are **ELECTRICAL**, and it takes time for them to zap through your neurons. The speed is different for different neurons.

Scientists have found that it takes longer for people to react to a shock to their toe than a shock to the bottom of their **SPINE**. This is because the message has to travel farther to get from the toe to the brain.

Messages can be as slow as 1 mile (1.6 km) per hour . . .

. . . or as fast as 267 miles (430 km) per hour. That's faster than a race car.

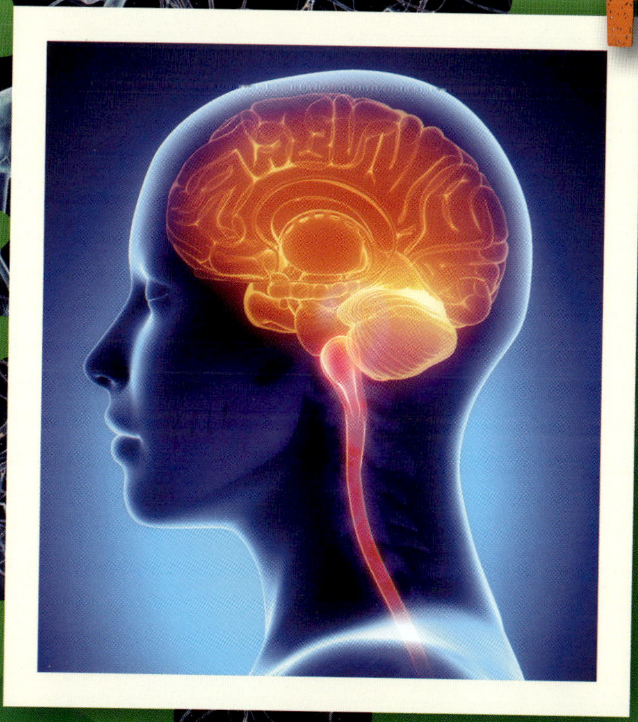

RACE to Calculate

Human brains and computers are similar in some ways. They both have electrical messages zooming along complicated networks. But which can think faster? It depends on what you measure. When it comes to basic math, humans are around ten million times slower than a computer. However, we are much better at other things, such as solving complex problems and being creative.

SUPER Strength

Hysterical STRENGTH

In 1982, Angela Cavallo found her son **UNCONSCIOUS** and trapped underneath a car. He had been working underneath the car when the jacks supporting it had collapsed. Cavallo showed enormous and superhuman strength when she lifted the car long enough for her neighbors to pull her son out. Scientists call this hysterical strength.

FIGHT or Flight

How did Cavallo lift that car? Was she bitten by a radioactive spider? Of course not. It is believed that this superhuman strength comes from something called the fight-or-flight response. When people are under stress or in danger, the body releases extra **ADRENALINE**. This extra dose of adrenaline increases blood pressure, heart rate, and breathing rate, allowing the body to work harder and faster.

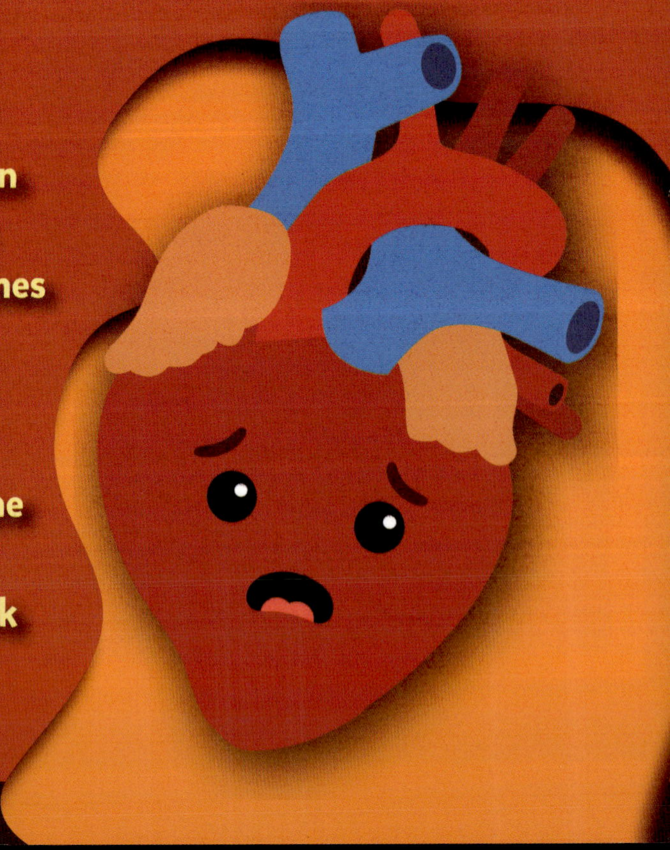

Is It Just TALK?

The evidence for hysterical strength isn't very strong. Although there are many examples of it, they are mostly **ANECDOTAL**. It is a very difficult thing to test because you would need to put someone into that fight-or-flight response, and for that to happen, the person must believe that the danger is real. So, unless scientists start acting like supervillains, we can't really test it.

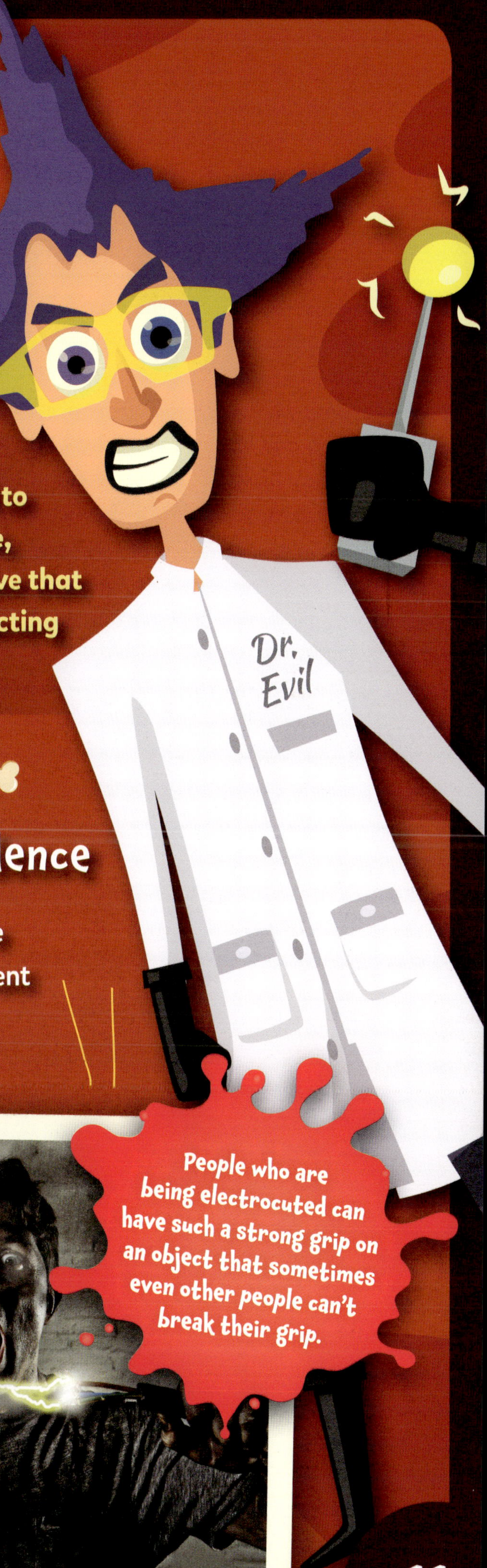

The ELECTRIC Evidence

However, there is some evidence that muscles are stronger than we thought. When an electric current is passed through a muscle, the muscle's **CONTRACTION** is much stronger than usual.

DANGER

Electricity is very dangerous and will not give you superpowers!

People who are being electrocuted can have such a strong grip on an object that sometimes even other people can't break their grip.

Holding Your

BREATH

for Fun

HOUDINI Put to Shame

How long can you hold your breath? For thirty seconds? For one minute? Harry Houdini, the famous escape artist of the early twentieth century, wowed crowds by holding his breath for over three minutes. In 2016, Aleix Segura held his breath for twenty-four minutes and three seconds!

Houdini was an amateur.

WHY Do We Need to Breathe?

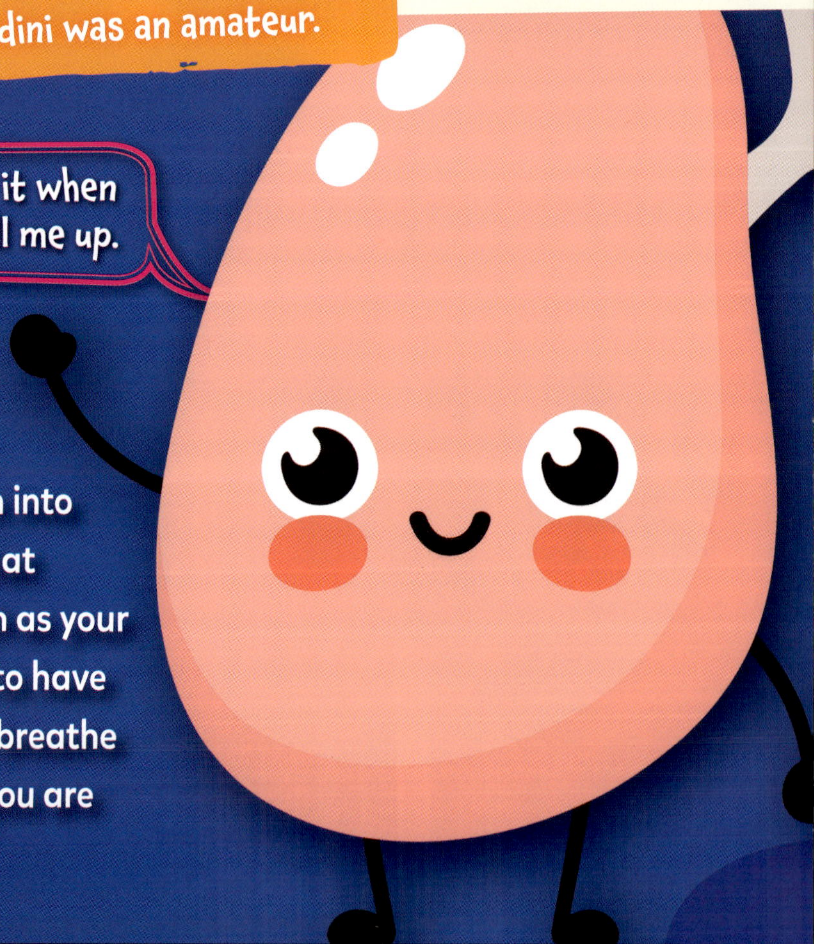

When you breathe, you take oxygen into your body. Your blood then takes that oxygen to all your vital organs, such as your heart and brain. Your organs need to have oxygen in order to work. When you breathe in deeply and hold your breath, you are storing oxygen to use later.

GASPING for Breath

Normally, if you are holding your breath, your body uses up the oxygen and produces **CARBON DIOXIDE (CO_2)**. This starts to build up in the lungs. It is not the lack of oxygen that causes us to desperately want to breathe—it is the buildup of CO_2.

Most people can hold their breath for around thirty seconds.

HOW Does Aleix Do It?

When Aleix holds his breath, he does it underwater. Strangely, our bodies work differently when we hold our breath underwater. Our organs go into a kind of energy-saving mode where they use less oxygen.

New YOU

The Same BIKE?

Imagine that you have the same bike for five years. First, the tires pop, so you replace them. Then the wheels bend, and you replace them. Before you know it, every part of the bike has been replaced. Is it the same bike?

This same question was asked over two thousand years ago, but then it was a boat and not a bike.

Are YOU a Bike?

Your body is made up of tiny building blocks called cells. When the cells in your body wear out, they are replaced by new cells, just as the parts of a bicycle can be replaced. So, are you just like a bike? Well, not quite.

The SAME You?

So, like your bicycle, all the parts have been replaced, but just like your bicycle, the rider is still the same. Some of your cells last for your entire life, such as the cells in your NERVOUS SYSTEM, your brain, and some of the cells in your eyes. It makes you wonder what we mean when we say "me" or "I."

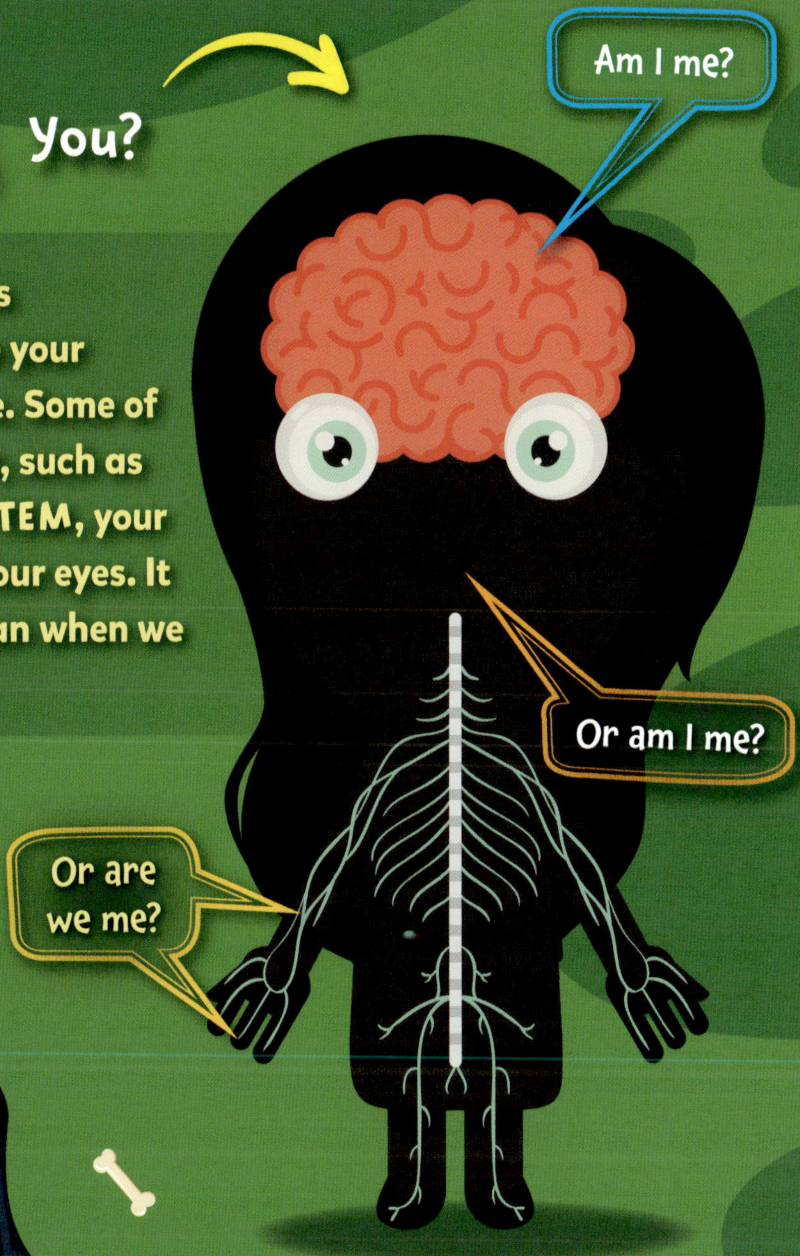

Frankenstein's MONSTER

When Dr. Frankenstein made his monster, do you think he had to replace the parts like our bicycle? Do you think he would say, "I have had the same monster for ten years. I gave it a new head, torso, brain, legs, and arms, but it is definitely the same monster"?

Land of the GIANTS

You are a stupid brain. I am much smarter than you!

So, You Think You're Pretty SMART, eh?

Brains like you to think they are very clever, don't they? I mean, that is where all the thinking takes place, and what is smarter than thinking? Well, it turns out that brains are pretty easy to trick if you know how, and some scientists in Sweden know exactly how to trick your brain.

BIG or Small

These scientists have found a way to trick your brain into thinking that you are either a giant or very small. The trick is to show the **TEST SUBJECT** an image of either a large or small version of the subject's body using a **MANNEQUIN** and poking it with a stick. OK, that does sound silly.

HOW It Works

The subjects wear a video display, a bit like a VR headset, so when they look down at their own bodies, they see the bodies and legs of a mannequin instead of their own. This mannequin body is prodded with a stick at the same time as the test subject.

SIZE Is Relative

The prodding technique tricks the brain into thinking that the tiny body is theirs. People reported seeing other people of normal size while wearing the headsets but thinking that they were giants. This seems very silly indeed, but it does teach us something about how our brains figure out the sizes of things around us. Maybe our brains decide how big we are by what we see and not what we already know.

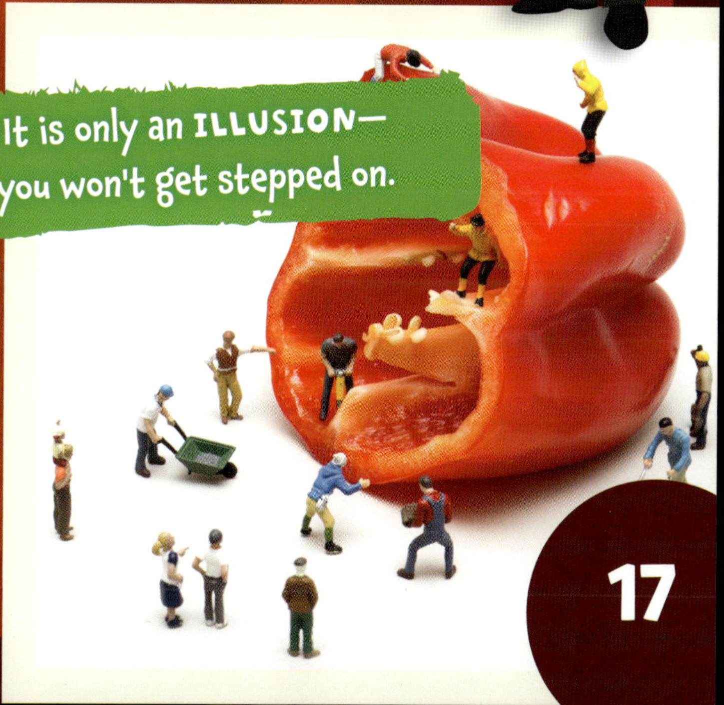

Fingernails Are GROSS

Nail BITING

Our fingernails are tough. Our fingernails are strong. Our fingernails are . . . dead? That's pretty gross. Think about that next time you want to bite your nails. But the weirdest thing about fingernails is that they never stop growing for the whole time that you are alive.

Your fingernails grow around three times faster than your toenails.

RECORD Breakers

The record for the longest fingernails was held by a man called Shridhar Chillal. Put together, his nails were longer than a bus. The longest nail was on his thumb and measured a massive 78 inches (198 cm)—if it was laid flat, it would've been longer than most grown-ups! Chillal cut off his nails. Well actually, he needed help from a nail technician and some power tools.

Fingernails grow more slowly as we get older. So, you better start now if you want to beat that record, but maybe ask your parents first!

NAILING the Science

Your fingernails are made of keratin, which is the same stuff that your hair is made of. Behind the cuticle on each finger and toe, your body makes a never-ending amount of living keratin. As the keratin gets pushed out of your finger, it hardens and dies and becomes your nail.

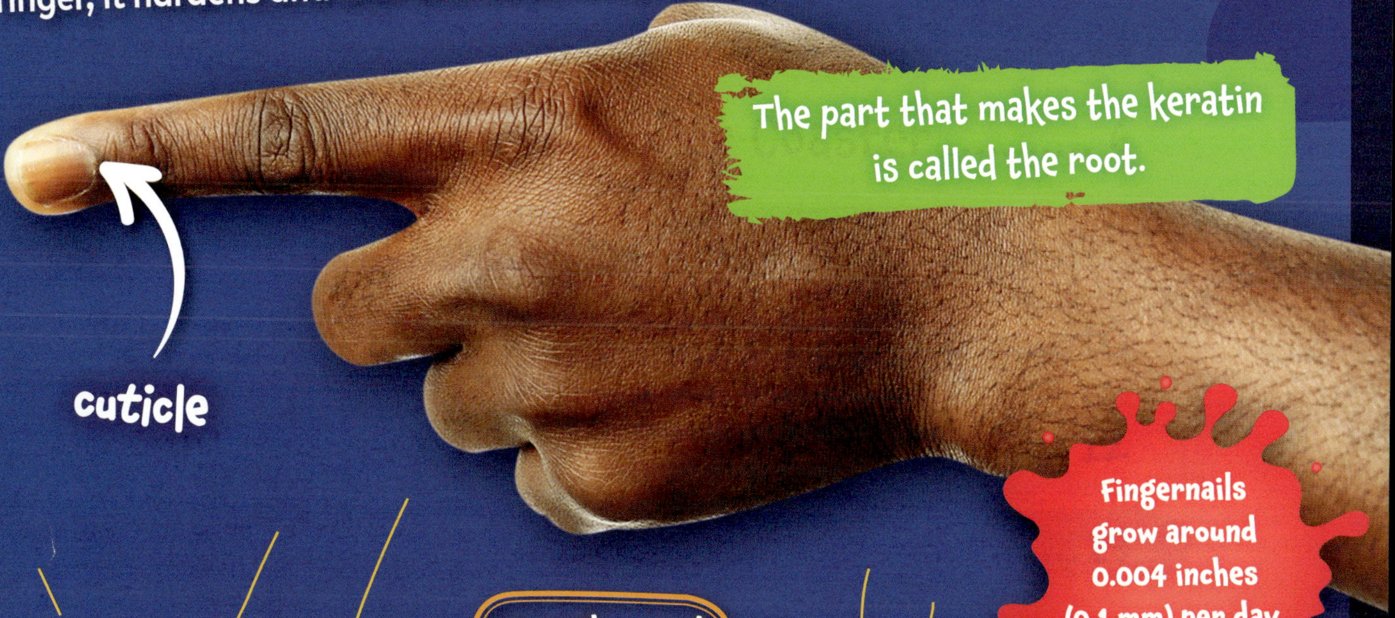

The part that makes the keratin is called the root.

Fingernails grow around 0.004 inches (0.1 mm) per day.

Life BEYOND Death

If you are very brave, you might have seen a horror movie about zombies. Zombies usually have long nails. This is because many people say that nails keep growing even when someone is dead. However, this is not true. When someone dies, the skin around the fingernails shrinks away. This makes it look like the nails have grown.

Nails growing after death? That's the silliest idea in this whole book.

How Many HICCUPS?

HIC ... Hic ...

Have you ever . . . hic . . . Sorry about that. Have you ever . . . hic, hic . . . Oh, bother! Have you ever had hiccups before? Isn't it annoying? You can't get a sentence out without interrupting yourself. But at least it usually passes after a while. Well, for one man, it didn't pass for a very long time indeed . . .

Charles OSBORNE

Charles Osborne had hiccups for sixty-eight years! Osborne hiccupped around once every three seconds when he was awake. Osborne says that he was working on his farm when he fell and knocked himself unconscious. When he woke, the hiccups began!

It is thought that he hiccupped over twenty-four thousand times per day. That's over five hundred million times in his lifetime!

WHAT Exactly Are Hiccups?

Just underneath your lungs is something called the diaphragm. The diaphragm is a muscle that controls your breathing. When it gets irritated, it **SPASMS**. This pulls air into the throat suddenly. When the air hits the back of your throat, your vocal cords suddenly close, making the famous hic sound.

I know those hiccups are in here somewhere!

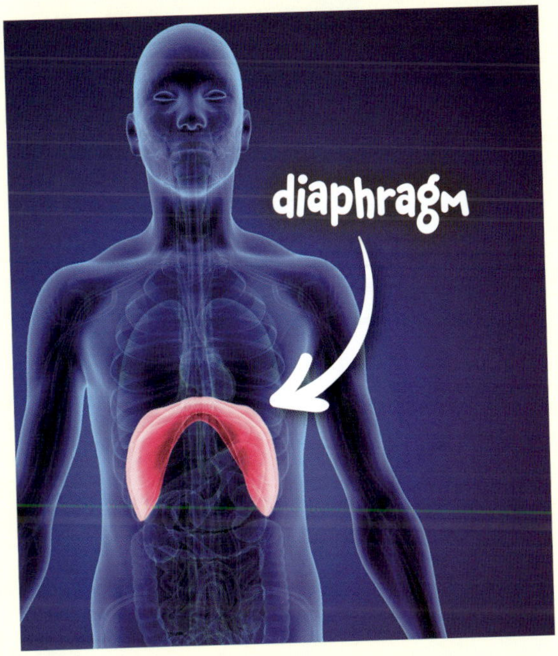

Can You CURE Hiccups?

Everyone knows a remedy for hiccups, be it drinking a glass of water while standing on your head, holding your breath until you are blue in the face, or having a friend frighten you. One of Osborne's friends tried to scare the hiccups out of him by shooting a shotgun behind him. It didn't work.

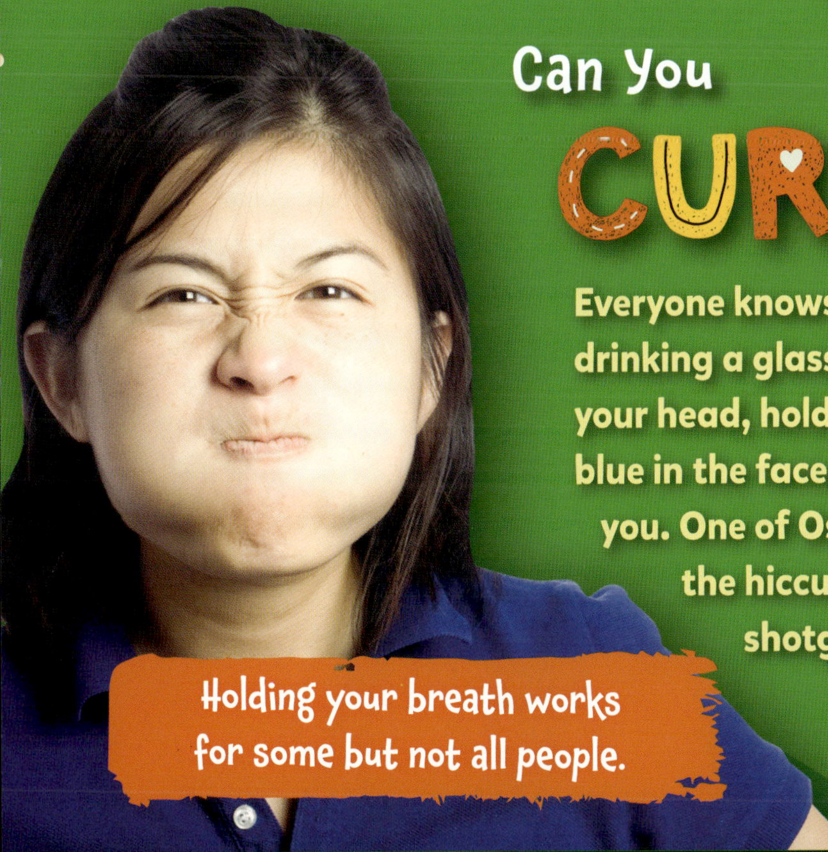

Holding your breath works for some but not all people.

RAPUNZEL,

Rapunzel, Let Down Your . . .

. . . HAIR?

Did it ever seem silly to you that Rapunzel would have visitors climb up her hair? Not only would that have been very painful for Rapunzel, but hair is probably quite hard to climb too. A clever visitor would have called, "Rapunzel, Rapunzel, let down your rope ladder!"

STRONG Hair

Human hair is surprisingly strong. In fact, if a strand of hair were as thick as a steel bar, it would be almost as strong! So, it isn't surprising that Rapunzel used her hair to pull up her visitors. But exactly how strong would her hair really be?

HOW

Strong Is Hair?

A single strand of hair is strong enough to hold around 3 ounces (85 g) without breaking. That is about the weight of two standard hot dogs. The average head holds around 150,000 hairs, so together they could hold three hundred thousand hot dogs!

REALLY?

How Strong Is Hair, Really?

So, if a single strand of hair can hold around 3 ounces, then all 150,000 hairs could hold around 13 tons (12 t)! That is about the weight of two African elephants. Good luck—I hope your neck muscles are strong!

PINKY Power

Secret WEAPON

Introducing the super finger!

You like holding things, don't you? What about climbing? Or tugs-of-war? Well, the secret weapon in all of those activities is the superpower of the pinky finger (and the ring finger). Is it a worm? Is it a cocktail sausage? No, it's the pinky finger.

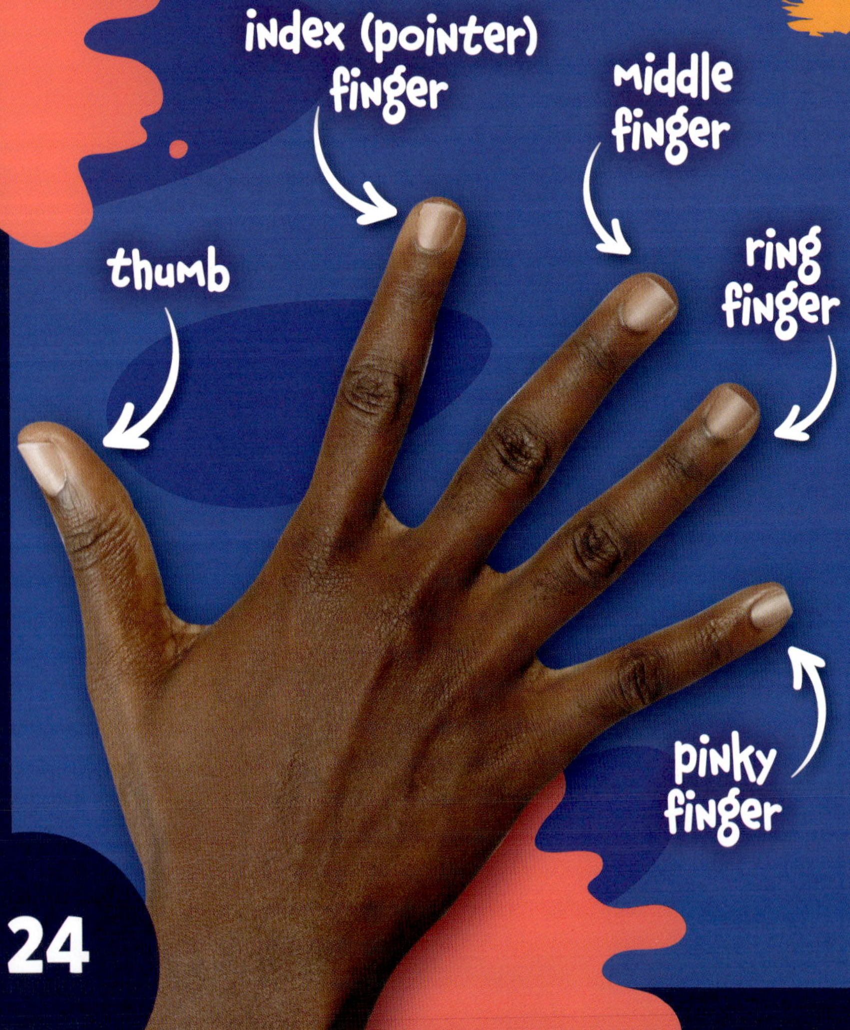

Mapping the HAND

You probably think the most important finger on your hand is the index (or pointer) finger because you need it to make finger guns. Even though your index finger and middle finger are the most **DEXTEROUS**, most of your strength comes from the ring finger and pinky finger.

GRIP Strength

Scientists tested people's strength when they could use all their fingers and then again when they couldn't use their little finger and their ring finger. The results were very surprising. Grip strength dropped by over half!

The Strength TEST

If you want to test this yourself, see how long you can hang from a set of monkey bars. Have a rest and then try it again using only your thumb, index finger, and middle finger to hold on.

The Hills Are Alive with the TASTE of Music

Ahh, smells like the number 4.

It Can Be a Confusing WORLD

You open your bedroom window and breathe in the sunlight; you taste the sound of the birds singing and hear the bright blue of the clear sky. Wait, is that right? It sounds like you have synesthesia.

WHAT Is Synesthesia?

These cakes smell very purple.

Synesthesia is an incredible condition where SENSES become confused. There are lots of different versions. For some people, when they smell something, they also have a color appear in their mind. For other people, they may get a sense of smelling something when they hear music.

The Taste of WORDS

Things such as numbers, letters, and words can be affected by synesthesia. For it to be synesthesia, the experience must be the same every time. If you smell a bad smell every time you hear a bad word, then it would be synesthesia. If it only happens sometimes, then it's more likely that someone just farted!

I don't even have synesthesia. . . . Did you just . . .

Some types of synesthesia are more uncommon than others. One woman experiences different places, objects, and people around her as being different times.

There Is PROOF

In the early 1990s, Professor Simon Baron-Cohen asked people with synesthesia what color they saw when he gave them a letter or number. Even when asked months or years later, people with synesthesia gave the same answers.

World's Loudest BURP

To Burp Is to Be HUMAN

Some things about being a human are unavoidable. These are the things that connect us and make us not so different from one another. One of those things is gas. It can come out of either end, but one thing is for sure: to burp is to be human!

The Disgusting SCIENCE of Burping

Why do you burp? It's gas! When you eat and drink, you take in more than just food—you are also swallowing gas. This gas needs to go somewhere. More than that, some foods make extra gases inside your stomach. When the gas builds up, it has to escape.

The World's LOUDEST Burp

A man from Bognor Regis in the United Kingdom holds the record for having the loudest burp in the world! It was recorded at 109.9 DECIBELS (dB). That is louder than a toilet flushing at 75 dB, riding a motorbike at 100 dB, and is about as loud as the average rock concert at between 103 and 110 dB!

sound level meter

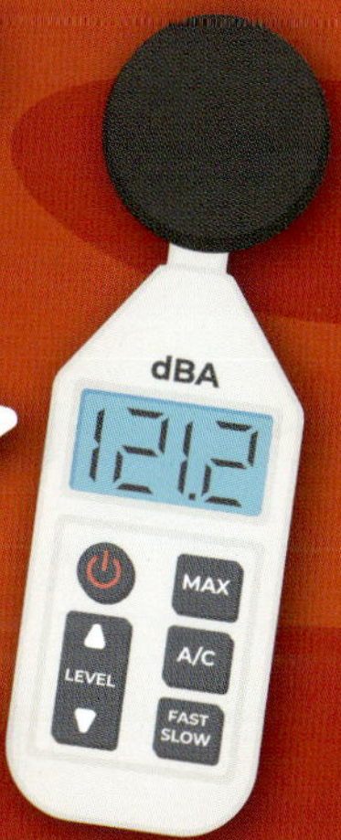

Dangerous BURP

Believe it or not, that record-breaking burp is loud enough to damage a person's ears. Sounds that are above 85 dB can damage people's hearing if they are around that sound for long enough!

BODIES Are Silly

From our superfast thoughts to our incredible lungs, from never-ending hiccups to superstrong hair, if there is one thing that we have learned about bodies, it is that they are silly.

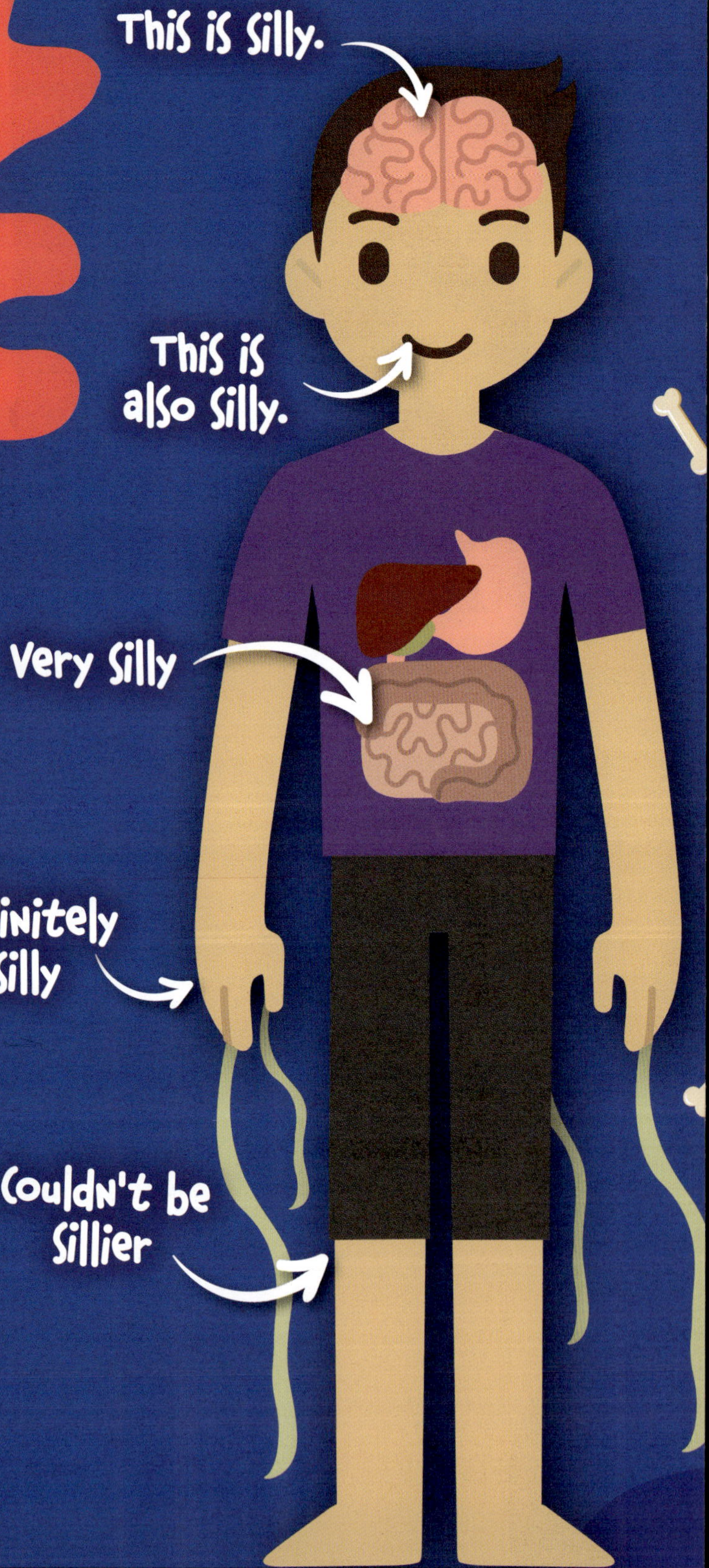

GLOSSARY

adrenaline	a chemical released in the body that raises heart rate and blood pressure, and gets muscles ready to exercise
anecdotal	known through stories that cannot be proven or shown to be true
carbon dioxide (CO_2)	a natural, colorless gas that is found in the air
cells	the basic units that make up all living things
contraction	when muscles squeeze together to become shorter and tighter
decibels	the units used to measure how loud a sound is
dexterous	being very skilled, especially with your hands
electrical	to do with creating or controlling electricity
gravity	the force that pulls everything down toward the center of Earth
illusion	something that appears to be real or true but is not
mannequin	a human-shaped dummy or doll
nervous system	the system of nerves and cells that signals from the body and brain travel along
network	a group of systems that are connected
orbit	the path that something follows around bigger objects in space
senses	things such as sight, smell, taste, and touch that help us experience the world
spasms	when muscles move uncontrollably
spine	the series of bones that run up the back of the neck and support the throat
test subject	someone who takes part in a scientific experiment
unconscious	a state of being without awareness, such as sleep

INDEX

PHOTO CREDITS: Images are courtesy of Shutterstock.com.
With thanks to Getty Images, Thinkstock Photo, and iStockphoto.

Cover: Droidworker, Recurring Images: bosotochka (header font), Sonechko57 (blobs), Rhenzy (bone vectors), Amy Li (additional illustrations). P2-3 - Tsvetkov Maxim, p4-5 - Elnur, Dmitry Lobanov, NotionPic, svtdesign, Rachael Arnott, Tartila, Sunspire, p6-7 - Yindee, AnastasiaN, Droidworker, svtdesign, PremiumArt, VIAVAL, p8-9 - CLIPAREA l Custom media, Daxiao Productions, whitehoune, HomeArt, MaryValery, p10-11 - Mooi Design, ProStockStudio, Yindee, Marcos Mesa Sam Wordley, Puslatronik, p12-13 - Baksiabat, svtdesign, ann13131, JonMilnes, p14-15 - ProStockStudio, Jacek Chabraszewski, oo23, Beatriz Gascon J, Natee K Jindakum, p16-17 - Stocklifemax, canbedone, feeling lucky, Ebtikar, JuliaGrin, Sean Voelger, Fourleaflover, p18-19 - tynyuk, Africa Studio, Oxy_gen, pathdoc, Vectorpocket, p20-21 - Daxiao Productions, Amanita, Silvicora, paper_Owl, sciencepics, Lucky Business, ann131313, p22-23 - Asier Romero, Iryna Kuznetsova, NotionPic, Korosi Francois-Zoltan, grmac, Sujono sujono, p24-25 - Matthias G. Ziegler, Lorelyn Medina, Africa Studio, p26-27 - VH-studio, pathdoc, Olena Chukhil, Inspiring, Iconic Bestiary, p28-29 - Jethra Tull, SPS Media, Panda Vector, svtdesign, Goran Djukanovic, son Photo, ZoriDj, p30 - sudowoodo.